AQA

AS/A LEVEL

WORKBOOK

Computer Science 1

Mark Clarkson

HODDER EDUCATION
LEARN MORE

Contents

WORKBOOK

(1) This workbook will help you to prepare for the **AQA Computer Science Paper 1** exam.

(2) Your **exam** is 2 hours 30 minutes and includes a range of questions. The exam is divided into four sections. In Section A you will answer questions about data structures, computation and algorithms. In Section B you will write a computer program to carry out a given algorithm. In Section C you will answer questions about the skeleton code provided by AQA. In Section D you will add extra functionality to the skeleton code provided by AQA.

(3) For **each topic** of Paper 1 there are:
- stimulus materials including key terms and concepts
- short-answer questions that build up to exam-style questions
- space for you to write or plan your answers.

(4) **Answering the questions** will help you to build your skills and meet the assessment objectives AO1 (knowledge and understanding), AO2 (application) and AO3 (design, program and evaluate computer systems).

(5) You still need to read your textbook and refer to your revision guides and lesson notes.

(6) If you are an A-level student you should answer all questions in the workbook. If you are an AS student you should answer only those questions in the lighter tint. Content specific to A-level is indicated by a red line in the margin.

(7) Because the skeleton code changes every year, this workbook is designed to help you develop and practise the relevant skills that will help you, but does not include specific information relating to each year's scenario.

(8) **Timings** are given for the exam-style questions to make your practice as realistic as possible.

(9) **Marks** available are indicated for all questions so that you can gauge the level of detail required in your answers.

(10) Answers are available at:
www.hoddereducation.co.uk/workbookanswers

Fundamentals of programming

Programming

The programming concepts assessed in this exam are not specific to any particular language. You should be particularly familiar with one language (C#, Java, Pascal/Delphi, Python or VB.net) in order to tackle the practical elements of the assessment and you should be confident in applying each of the techniques in this topic in that language.

Programming concepts

A key concept in computer programming is how values are stored and referred to using variables. The concept of a data type, and that different types of variables are needed to store different types of data, is a fundamental building block for understanding and creating computer programs. You should be familiar with using numeric data types (integer, float/real), text data types (string, char), Booleans and simple data structures including arrays. You should also know that a value can be stored as a named constant.

The main programming skills can be broken down into six constructs: variable declaration, constant declaration, assignment, iteration (both condition-controlled and count-controlled loops), selection (if statements) and subroutines (procedures and functions). There will be questions about how these concepts are applied in sections A and C, and you will definitely be expected to use these concepts to create your own programs and subroutines for sections B and D. You will almost certainly have to complete a dry-run or trace table to show that you can follow a program using these concepts.

Operations

You are expected to be able to carry out various operations within the code that you write. The syntax for this will be specific to your centre's chosen programming language and the mark schemes are different in each case to allow for this.

- Arithmetic operations include the basic addition, subtraction and multiplication as well as real/float division, integer division and remainders. You are also expected to be able to use powers (exponentiation), rounding and truncation (cut off the decimal portion of a real/float).
- Boolean operations such as AND, OR, NOT and XOR can be used to combine conditions in selection and condition-controlled loops.
- String-handling operations include finding the length of a string, the position of a character within a string, capturing a substring, concatenating (joining) strings and also converting the characters from a string to their ordinal (numeric) values and vice versa.
- Generating random numbers within a given range is a key skill you must make sure you are familiar with.
- Exception-handling syntax varies according to programming language, but most have a try ... catch structure for identifying and dealing with exceptions, such as a failed attempt at casting.
- Casting is not explicitly named in the specification, but you must be confident in converting between various data types and strings.

Subroutines

Most computer programs are decomposed into subroutines known as procedures (which don't have a return value) and functions (which do). It is important to understand how to pass data to subroutines using parameters and also the effects of using local and global variables.

Advanced programming concepts

More advanced topics in this area include the use of stack frames, which keep track of the return address, parameters and local variables for each subroutine as it is called, and the use of recursive techniques (including general and base cases).

1 State the most appropriate data type for each of the following. (AO1) **5 marks**

a 12.76

....

b 'Tea'

....

c 12:06

....

d −17

....

e false

....

2 A program is written for calculations involving circles. The programmer has decided to make use of a named constant. (AO1, AO2) **3 marks**

a State the definition of a constant.

....

b State one advantage for using a named constant for a constant value.

....

....

3 Figure 1 shows part of an algorithm. (AO1) **4 marks**

```
IF Score > 60 THEN
    PRINT "Well done!"
ELSEIF Score >= 30 THEN
    PRINT "Not too bad"
ELSE PRINT "Oh dear…"
ENDIF
```

Figure 1

Complete the table to show the message that would be displayed for each value of Score.

Score	Message displayed
25	
60	
30	
100	

4. Programs can contain two different types of loops – condition-controlled loops and count-controlled loops. (AO1)

a. Which term above describes a WHILE loop?

....

b. Which term above describes a FOR loop?

....

c. Why would a programmer use a condition-controlled loop over a count-controlled loop?

....

5. Figure 2 shows part of an algorithm. (AO1, AO2)

```
FUNCTION CalcVolume(Length, Width, Height)
   Area ← Length * Width
   Volume ← Area * Height
   RETURN Volume
ENDFUNCTION
```

Figure 2

Figure 3 shows part of a program that calls the algorithm in Figure 2.

```
Area ← 0
Volume ← CalcVolume (3,4,5)
```

Figure 3

a. State the number of parameters CalcVolume takes.

....

b. State the identifier for a parameter of CalcVolume.

....

c. State the identifier for a local variable within CalcVolume.

....

d. When the section of program in Figure 3 has finished, state the value of the variable Area, and explain why it has this value.

....

....

....

e. State two reasons why it is usually preferable that subroutines don't use global variables.

....

....

f. State the difference between a function and a procedure.

....

....

6 Figure 4 shows part of an algorithm. (AO2)

```
FUNCTION QuestionSix(Number)
    FOR f ← 2 TO Number - 1 DO
        IF Number MOD f = 0 THEN
            RETURN False
        ENDIF
    ENDFOR
    RETURN True
ENDFUNCTION
```

The MOD operator calculates the remainder from an integer division, e.g. 15 MOD 2 = 1.

Figure 4

a   Complete each of the dry run tables below. You may not need to use all of the rows.

i

Number	f	Return statement
5		

ii

Number	f	Return statement
25		

b   State the purpose of this function.

7 Figure 5 shows part of an algorithm for displaying the numbers in the Fibonacci Sequence. (AO1, AO2)

```
FUNCTION Fib(n)
    IF n <= 2 THEN
        Return 1
    ELSE
        x ← Fib(n-1)
        y ← Fib(n-2)
        RETURN x + y
    ENDIF
ENDFUNCTION
```

Figure 5

a   Fib is a recursive subroutine. Explain what is meant by a recursive subroutine.

b   State the base case for the subroutine Fib.

c Complete the table below to show the result of tracing the Fib algorithm. You may not need to use all of the rows.

Call number	n	x	y	Value returned
1	4			
2	3			
3	2			1
2	3	1		
4				
2				
1				
5				
1				

d State two items that will be stored in the stack frame when the Fib subroutine is called.

....

e State the largest number of stack frames at any given point when the subroutine call Fib(4) is made. State which call numbers will be on the stack for each case.

....

Exam-style questions

(25)

8 Figure 6 shows part of an algorithm for recording scores in a cricket match. (AO1, AO2) (6 marks)

```
1  FUNCTION CricketScore()
2     Runs ← 0
3     Overs ← 0
4     Batter ← 11
5     MAX_OVERS ← 20
6     WHILE Batter > 1 AND Overs < MAX_OVERS
7        Balls ← 0
8        FOR Balls ← 1 TO 6 DO
9           Out ← INPUT "Is the player out?"
10          IF Out = "Yes" THEN
11             Batter ← Batter - 1
12          ELSE
13             Score ← INPUT "Enter score"
14             Runs ← Runs + Score
15          ENDIF
16       ENDFOR
17       Overs ← Overs + 1
18    ENDWHILE
19 RETURN Runs
20 ENDFUNCTION
```

Figure 6

a   From the program above, state the line number of an assignment operation.

....

b   State the identifier of a named constant.

....

c   State the identifier of an integer.

....

d   State the line number of a selection structure from the program above.

....

e   State the line number of a condition-controlled loop from the program above.

....

f   Explain why a condition-controlled loop has been used in this case.

....

....

9 The contents of the array Binary are shown in Figure 7.

The algorithm, represented using pseudo-code in Figure 8, describes a method to convert the denary (base 10) representation of a number into a binary (base 2) representation of the same number, using the array Binary to store the end result. (AO2, AO3)

11 marks

Binary			
[0]	[1]	[2]	[3]
0	0	0	0

Figure 7

```
Binary[4] ← [0,0,0,0]

Denary ← INPUT "Enter a whole number between 0 and 15: "

BITS ← 4

FOR i ← 0 TO BITS-1

    Pos ← BITS - 1 - i

    Binary [Pos] ← Denary MOD 2

    Denary ← Denary DIV 2

ENDFOR

FOR i ← 0 to BITS-1

    PRINT Binary[i]

ENDFOR
```

Figure 8

The MOD operator calculates the remainder result from an integer division, e.g. 15 MOD 2 = 1

The DIV operator calculates the integer division, e.g. 15 DIV 2 = 7

a   Write a program for the algorithm in Figure 8.

(Note: This should be done on a computer, using the programming language you will be using to carry out your Paper 1 exam. You may choose to write out the code in the space below.)

b   Test the program by showing the result of entering 13 when prompted by the program.

c   Exception handling is a way of responding to fatal runtime errors. Describe how exception handling could be used in the algorithm in Figure 8.

Programming paradigms

The two major programming paradigms to consider are procedural and object-oriented programming (OOP).

In procedural programming there is typically one program that uses procedures and functions to break the various tasks up into modular sections. These are sometimes shown using a hierarchy chart, where a link is created to show each subroutine call. Questions involving hierarchy charts are typically asked with reference to the skeleton code.

In object-oriented programming (OOP) the problem to be solved is split into classes, where each class describes a type of real-world entity, e.g. a Tiger class is a blueprint for all tigers. The class definition includes attributes (the variables that relate to each tiger – height, weight, gender) and methods (subroutines – run, walk, grow).

Each instance of a tiger is instantiated as an object and has its own values for each attribute (i.e. two different tigers may have different heights and weights). This paradigm more closely models the real world, though it is more complex to program.

Inheritance and composition are key aspects of object orientation to describe the way that different objects can often use the same methods in order to reduce the duplication of code.

1 Describe two advantages for using procedures and functions in a computer program. (AO1)

4 marks

....

....

....

....

....

2 Figure 9 shows an incomplete hierarchy chart for a game of Rock, Paper, Scissors. (AO1, AO2)

4 marks

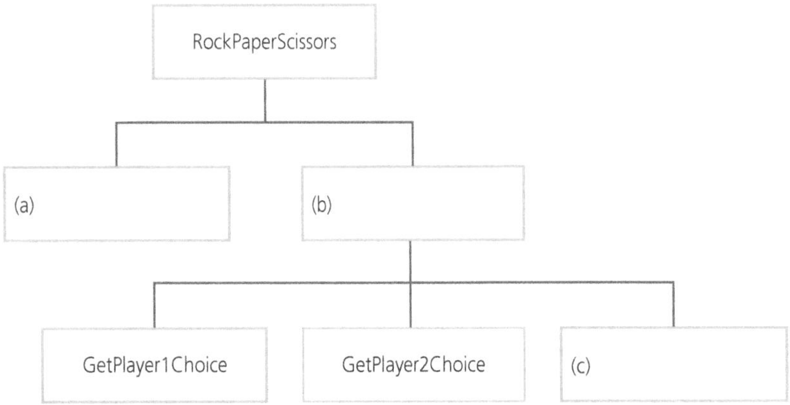

Figure 9

The program contains the following subroutines:

```
DecideWinner

DisplayRules

GetPlayer1Choice

GetPlayer2Choice

PlayOneRound

RockPaperScissors
```

a   What should be written in box (a) in Figure 9?

....

b   What should be written in box (b) in Figure 9?

....

c   What should be written in box (c) in Figure 9?

....

d   A procedural programming approach has been used in the development of the program described in Figure 9. Explain what is meant by a procedural programming approach.

....

3 Figure 10 shows a class diagram for a program to model various types of vehicle. (AO1, AO2)

15 marks

Car
–Make: String
–Model: String
–EngineSize: Float
–FuelType: String
–FuelLevel: Float
+Car()
~GetFuelLevel()
+Drive()
+Park()
-ApplyHandbrake()
-ReleaseHandbrake()

Figure 10

a   State the identifier for one attribute.

....

b   State the identifier for one public method.

....

c   State the identifier for one private method.

....

d   State the identifier for one protected method.

....

e   State the name of the constructor for this class and explain why this must be the case.

....

....

f   Explain what is meant by the term 'instantiation'.

....

....

g   Describe the relationship between a class and an object of that class.

....

....

h   Explain the difference between public, private and protected methods.

....

....

....

i   There is an attribute FuelLevel. Explain why there is also a method GetFuelLevel.

....

....

j Explain why this approach is favoured in object-oriented programming.

....

4 Another programmer has suggested that the program should be designed using inheritance and has created the class diagram in Figure 11. (AO1, AO2) **5 marks**

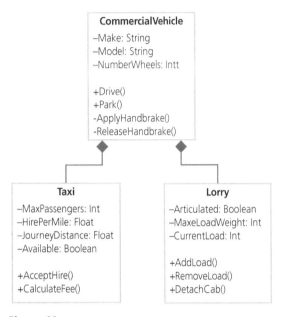

Figure 11

a Explain what is meant by inheritance.

....

b State the identifier for a base class shown in Figure 11.

....

c State the identifier for a subclass shown in Figure 11.

....

d Explain why inheritance is preferable in this situation.

....

....

5 Part of the class definition for Bicycle has been represented in Figure 12. (AO3) **4 marks**

```
Bicycle = Class(Vehicle)

    Private:

        NoOfGears: Int

        BrakeType: String

        CurrentGear: Int

    Public:

        Procedure: ChangeUp()

        Procedure: ChangeDown()

        Function: GetCurrentGear()
```

Figure 12

A new vehicle type is to be added to the program, called ECycle. The class ECycle is to be a subclass of the Bicycle class. When an ECycle is inspected it should display all the information shown for a normal bicycle plus the additional information stored about an ECycle.

An ECycle has the following additional attributes:

■ MaxCharge: Stores the maximum charge, in Amp Hours (AH), that can be stored in the battery.
■ CurrentCharge: Stores the current charge, in Amp Hours (AH).

An ECycle has the following additional methods:

■ FullyCharged(): Returns True if the CurrentCharge is equal to the MaxCharge.
■ IsEmpty(): Returns True if the CurrentCharge is 0.

Write the class definition for ECycle, using similar notation to that used in Figure 12.

....

....

....

....

....

....

....

6 A wildlife modelling system is being designed and will contain a base class Animal followed by inherited classes Bird and Mammal. A class diagram is shown in Figure 13. (AO1, AO2)   **7 marks**

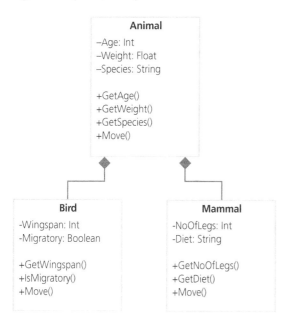

Figure 13

a   All three classes include a procedure with the identifier Move. State the name of the programming concept being used and describe how it functions.

....

....

....

b   Explain why this programming technique is appropriate in this case.

....

....

c   The design in Figure 13 is aligned with the principle 'Encapsulate what varies'. Explain, using examples from Figure 13, what this principle means.

....

....

....

7 A building management company are creating an object-oriented program to store details about the various buildings that they manage. A class Building has been created and a subclass Room is to be designed. (AO1, AO2, AO3)   **7 marks**

Part of the definition of the Building class is:

```
Building = Class
    Private:
        Type: String
        Town: String
    Public:
        Function: GetType
        Function: GetTown
        Procedure: SetDetails
```

A room has the following additional fields:

■ Windows: Stores True if the room has any windows at all or False if it is an internal room with no windows.
■ Rent: Stores the cost (per week) to rent this room.
■ Occupied: Stores True if the room is currently occupied or False if it is vacant.

a   Write the class definition for Room.

....

....

....

....

....

....

b   Describe the difference between association aggregation and composition aggregation.

....

....

c   Explain why the use of composition aggregation would be appropriate for the relationship between the Building class and the Room class.

....

Exam-style questions

8 Figure 14 shows an incomplete hierarchy chart for a sandwich-making robot. (AO1, AO2)

7 marks

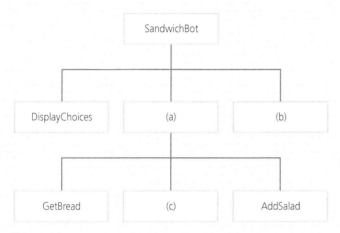

Figure 14

The program contains the following subroutines:

    AddMeat
    AddSalad
    DisplayChoices
    GetChoices
    GetBread
    MakeSandwich
    SandwichBot

a   What should be written in box (a) in Figure 14?

....

b   What should be written in box (b) in Figure 14?

....

c   What should be written in box (c) in Figure 14?

....

d   Describe how values can be passed between the different procedures in Figure 14.

....

....

....

e   Explain why it is preferable to use a procedural paradigm such as the one used in Figure 14 rather than designing and creating a program that does not use procedures.

....

....

....

....

9 A college requires a piece of software to organise the management of staff and student data. A class Person has been created and part of the implementation of that class is shown below. Two subclasses, Student and Teacher, are to be designed. (AO1, AO2, AO3) **11 marks**

```
Person = Class
    Private:
        FirstName: String
        Surname: String
    Public:
        Function: GetFirstName
        Function: GetSurname
        Procedure: SetupDetails
```

The subclass Teacher requires the following additional attributes:

- Subject: The name of their preferred subject specialism.
- Paygrade: A numeric figure from 1 to 20 to show their current position on the pay scale.

The subclass Student requires the following additional attributes:

- Tutor: An object of class Teacher that represents the teacher assigned to be that students' pastoral tutor.

The subclass Student requires the following additional methods:

- Enrol: a procedure that will allow a student to enrol on a new course.

a   Write the class definition for Student.

....

....

....

....

....

....

....

b   State whether the class structure described in this scenario would be better suited to composition aggregation or association aggregation.

....

c   Describe one similarity and one difference between the two forms of aggregation and explain why the alternative method would not be appropriate in this case.

....

....

....

d   Explain what is meant by 'overriding' and why it is necessary in this case.

....

....

Fundamentals of data structures

Data structures

The main data structures used are arrays and files.

Arrays

A one-dimensional array is a fixed-length set of elements of the same data type that can be addressed or indexed. A two-dimensional array can be drawn as a table structure, or interrogated by considering it as an array of arrays. Arrays can be of any number of dimensions.

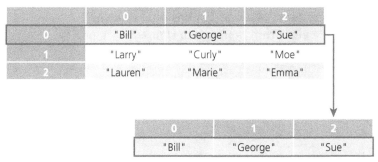

	0	1	2
0	"Bill"	"George"	"Sue"
1	"Larry"	"Curly"	"Moe"
2	"Lauren"	"Marie"	"Emma"

	0	1	2
	"Bill"	"George"	"Sue"

Figure 15

1 **Figure 16 shows the contents of a one-dimensional array of integers, Numbers. Figure 17 shows part of a pseudo-code algorithm. (AO1)**

Numbers

[0]	[1]	[2]	[3]	[4]
7	3	9	12	6

Figure 16

```
FUNCTION SumBigNumbers (Numbers)

    Total ← 0

    FOR i ← 0 TO 4 DO

      IF Numbers[i] > 5 THEN

          Total ← Total + Numbers[i]

      ENDIF

    ENDFOR

    PRINT Total

ENDFUNCTION
```

Figure 17

a **State the value of Numbers[2].**

....

b **State the index of the value 6.**

....

c Dry run the algorithm in Figure 17 and complete the table below.

(Note: It may not be necessary to complete all the rows.)

i	Total	Numbers				
		[0]	[1]	[2]	[3]	[4]
	1	7	3	9	12	6

2 The contents of a two-dimensional array of integers, R, are shown in Figure 18. The array is addressed so that the value of R[0,1] is 7. The contents of a one-dimensional array of integers, A, are shown in Figure 19. A pseudo-code algorithm is shown in Figure 20. (AO1, AO2)

9 marks

R				
	[0]	[1]	[2]	[3]
[0]	9	7	8	4
[1]	6	2	3	1
[2]	7	8	8	9

Figure 18

A		
[0]	[1]	[2]
0	0	0

Figure 19

```
FUNCTION QuestionTwo (R, A)
    FOR i ← 0 TO 2
        T ← 0
        FOR j ← 0 TO 3
            T ← T + R[i,j]
        ENDFOR
        A[i] ← T / 4
    ENDFOR
ENDFUNCTION
```

Figure 20

a State the value of R[2,1].

....

b Which of the following represents a valid position within the two-dimensional array R?

R[3,3]   R[3,2]   R[2,3]

....

c State the index of R which holds the value 1.

....

d Dry run the algorithm in Figure 20 and complete the table below.
(Note: It may not be necessary to complete all the rows.)

i	j	T	[0]	[1]	[2]
					A
			0	0	0

e Describe the purpose of the algorithm described in Figure 20.

....

....

3 A computer program is being written to store booking details for a boarding kennel.
The owners of the kennel need to be able to store the type of dog and have been
using the following codes: (AO1, AO2)

TD: Toy Dog (e.g. Chihuahua)
SD: Small Dog (e.g. Terrier)
MD: Medium Dog (e.g. Spaniel)
LD: Large Dog (e.g. Labrador)

a Explain why it may be an advantage to store the dog type as part of a binary file instead of a
text file.

....

....

....

b Give two potential disadvantages for storing the dog type as part of a binary file.

....

....

....

Exam-style questions

④ Figure 21 shows the contents of a one-dimensional array of integers, B. Figure 22 shows the pseudo-code for an algorithm. (AO1, AO2)

7 marks

[0]	[1]	[2]	[3]	[4]	[5]	[6]	[7]
1	0	0	1	1	1	0	0

Figure 21

```
T ← 0

D ← 128

FOR i ← 0 TO 7

    V ← B[i] * D

    T ← T + V

    D ← D / 2

ENDFOR
```

Figure 22

a   State the value of B[3].

....

b   Dry run the algorithm in Figure 22 and complete the table below.

T	D	i	V

c   State the purpose of the algorithm in Figure 22.

....

⑤ A computer program is being developed to simulate a board game that involves four players, each with an associated colour – Blue, Red, Green or Brown. Each tile on the board can be occupied by only one player at a time. (AO1, AO2)

4 marks

The following code is used to record the current state of each tile:

- U: Unowned
- Blue: Occupied by the blue player
- Red: Occupied by the red player
- Green: Occupied by the green player
- Brown: Occupied by the brown player

a   Describe a difference between the way data are stored in a text file and the way data are stored in a binary file.

....

....

b   Describe one advantage for storing the current state of the board in a text file.

....

c   Describe one advantage for storing the current state of the board in a binary file.

....

6   Figure 23 shows the contents of a two-dimensional array of integers, AM. Figure 24 shows the contents of a two-dimensional array of integers, AL. Figure 25 shows part of a pseudo-code algorithm. (AO1, AO2)   **8 marks**

AM	[1]	[2]	[3]	[4]
[1]	0	76	257	124
[2]	76	0	24	87
[3]	257	24	0	54
[4]	124	87	54	0

Figure 23

AL	[1]	[2]	[3]
[1]	0	0	0
[2]	0	0	0
[3]	0	0	0
[4]	0	0	0

Figure 24

```
FOR i ← 1 TO 4
    k ← 1
    FOR j ← 1 TO 4
        IF AM[i,j] > 0 AND AM[i,j] < 100 THEN
            AL[i,k] ← j
            k ← k + 1
            OUTPUT "T"
        ELSE
            OUTPUT "F"
        ENDIF
    ENDFOR
ENDFOR
```

Figure 25

Dry run the algorithm in Figure 25. Complete the table below.

i	j	k	Output

b Complete the table below to show the contents of the two-dimensional array of integers AL following the completion of the algorithm shown in Figure 25.

AL	[1]	[2]	[3]
[1]			
[2]			
[3]			
[4]			

Abstract data types

There are a wide range of abstract data types that can be used in a variety of programming situations. Some programming languages will have some abstract data types built in, but most can be recreated using an array (a static data type with a fixed size) or a list (a dynamic data type with a variable size).

Queues and stacks

Both queues and stacks can be easily modelled with one-dimensional arrays. Queues are First In First Out (FIFO) and behave like a real queue. Items are enqueued at the back of the queue and dequeued from the front.

Stacks are Last In First Out (LIFO) and behave like a stack of plates. Items are pushed on to the top of the stack and popped off when needed.

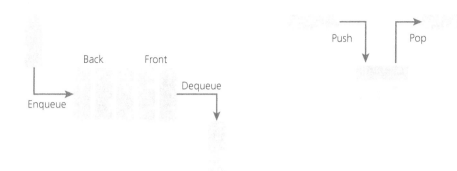

Figure 26

Graphs and trees

Graphs are made up of nodes and edges (or vertices and arcs). They can be represented graphically as an adjacency matrix or as an adjacency list. Graphs can be weighted or unweighted or directed or undirected. They are typically used to represent complex relationships. Trees are undirected, connected graphs that don't have any cycles. Binary trees have a maximum of two child nodes from any given node, and are often used to store sorted data.

Hash tables and dictionaries

Hash tables use algorithms to map each value in a list to a key, which can improve efficiency when searching through a database or can be used to encrypt passwords. Dictionaries use key–value pairs that allow a value to be accessed via the associated key.

Vectors

A vector can be represented as coordinates, an arrow, a dictionary, a list or a function. They are often used in physical simulations and can be operated on using:

- vector addition (finding the result of applying two or more vectors)
- scalar multiplication (finding the result of applying a vector multiple times)
- convex combination (finding a result between two or more vectors)
- scalar/dot product (used to find the angle between two vectors).

1 A programmer is unsure whether to use an array or a list. One is a static data structure and the other is dynamic. (AO1, AO2)　　　**4 marks**

　a　State the name of the data structure from the sentence above that is static.

　　　...

　b　State the name of the data structure from the sentence above that is dynamic.

　　　...

　c　Describe the difference between a static and a dynamic data structure.

　　　...

2 Transactions at a bank are placed into a queue for processing. (AO1, AO2)　　　**17 marks**

　a　Explain what is meant by a LIFO structure and a FIFO structure, and state which structure is being used in this case.

　　　...

　　　...

　　　...

b   The queue has been implemented using an array of 20 transactions. Transactions can be queued and dequeued from this array structure. Describe one possible error that could occur when using the queue structure and explain how to avoid this error.

....

....

....

c   Explain the main differences between a linear queue and a priority queue, describing why a priority queue may be preferable.

....

....

....

d   Describe the main differences between a linear queue and a circular queue, describing why a circular queue may be preferable.

....

....

....

e   Describe the main differences between a priority queue and a circular queue, describing a disadvantage for each.

....

....

3 The diagram shown in Figure 27 shows the train routes between five cities, labelled A–E. (AO1, AO2)

9 marks

Figure 27

a   The diagram in Figure 27 is an abstraction. Explain what is meant by the term 'abstraction'.

....

b   Tick one box that identifies this type of diagram.

Type of diagram	Tick one box only
Unconnected graph	
Directed graph	
Weighted graph	
Undirected graph	

c   Explain why the diagram in Figure 27 cannot be described as a tree.

....

d Complete the table below to represent the diagram in Figure 27 as an adjacency list.

A	
B	
C	
D	
E	

e Complete the table below to represent the diagram in Figure 27 as an adjacency matrix.

	A	B	C	D	E
A	0	0	0	1	0
B					
C					
D					
E					

f Describe the circumstances when it is more appropriate to represent a graph as an adjacency list rather than an adjacency matrix.

....

....

....

....

4 Figure 28 shows a two-dimensional array, M. Figure 29 shows a two-dimensional array, L. Figure 30 shows a pseudo-code representation of an algorithm. (AO1, AO2) **10 marks**

M	[1]	[2]	[3]	[4]	[5]
[1]	0	123	45	81	12
[2]	123	0	34	82	217
[3]	45	34	0	161	51
[4]	81	82	161	0	26
[5]	12	217	51	26	0

Figure 28

L	[1]	[2]	[3]	[4]	[5]
[1]					
[2]					
[3]					
[4]					
[5]					

Figure 29

```
ArrSize ← 5
FOR i ← 1 TO ArrSize
    FOR j ← 1 TO ArrSize
        IF M[i, j] < 100 AND IF M[i,j] <> 0 THEN
            Append(L, i, j)
        ENDIF
    ENDFOR
ENDFOR
```

Figure 30

The procedure Append takes three arguments: ArrayName, Record and Node. It will append the integer Node to the end of row Record of the two-dimensional array ArrayName.

For example, the call Append(L,1,2) would append the value 2 into the first empty space in row 1 of the array L.

a   Dry run the algorithm in Figure 30 by completing the table below. You may not need to use all of the rows provided in the table.

ArrSize	i	j	M[i,j]	Append (Y/N)
5	1	1		
```
```

b Complete the table below to show the contents of the two-dimensional array L after running the algorithm.

L	[1]	[2]	[3]	[4]	[5]
[1]					
[2]					
[3]					
[4]					
[5]					

c The two-dimensional array L represents an adjacency list for radio towers less than 100 km apart. Name two other suitable data structures that could be used to represent this information.

....

....

d Explain the purpose of representing the data held in the two-dimensional array M as an adjacency list.

....

....

....

....

....

5 Figure 31 shows four different graphs. (AO1, AO2)　　　　　　　　　　**11 marks**

 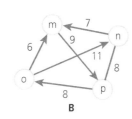  

　　　　A　　　　　　　　　B　　　　　　　　C　　　　　　　D

Figure 31

a State the letter of one graph that is a tree.

....

b State the letter of one graph that is unconnected.

....

c State the letter of one graph that is weighted.

....

d State the letter of one graph that is directed.

....

e   Describe two typical uses for a tree structure.

....

....

f   Explain what is meant by a rooted tree.

....

....

g   Explain what is meant by a binary tree.

....

h   Describe a typical use for a binary tree.

....

i   In graph B of Figure 31, which of the following would be an acceptable sequence?

m → p → o → p
n → o → p → m
n → m → p → o
m → p → n → o

....

6 A large company is implementing a hash table in order to store records relating to its employees. The hashing algorithm is described in Figure 32. Figure 33 shows the current state of the hash table. (AO1, AO2)   **12 marks**

The function Length(string) takes a single string as an argument and returns the total number of characters (excluding spaces) in that string.

The function VowelsIn(string) takes a single string as an argument and returns the total number of vowels in that string.

```
L = Length(name)

V = VowelsIn(name)

key = (L + V) MOD 10
```

Figure 32

Key	Record
0	
1	
2	
3	
4	
5	
6	
7	Anandna Sajid
8	Cameron Walker
9	

Figure 33

a   Calculate the key for an employee with the name Atif Hasan.

....

b   Calculate the key for an employee with the name Paula Jones.

....

c   A new employee called Christopher Way is to be added to the hash table, however the key generated is the same as that of Cameron Walker. Explain what the effect of this would be and suggest a possible solution.

....

....

....

d   A large number of names need to be added to the hash table so it has become necessary to rehash the data structure. Describe the steps required to complete the rehash.

....

....

....

e   Explain why a developer might choose to use a hashing algorithm to store data.

....

....

f   Explain how and why hashing algorithms are often used for storing passwords.

....

....

....

7 A sentence reads 'No sentence begins with because because because is a conjunction'. The sentence has been stored once in a dictionary called SentenceA and again in another dictionary called SentenceB. Both can be seen in Figure 34.   **5 marks**

SentenceA	
Key	Value
1	No
2	sentence
3	begins
4	with
5	because
6	is
7	a
8	conjunction

SentenceB	
Key	Value
No	1
sentence	1
begins	1
with	1
because	3
is	1
a	1
conjunction	1

Figure 34

a   State the value returned when looking up the key 6 in SentenceA.

....

b   State one use for the dictionary SentenceA.

....

c   State one use for the dictionary SentenceB.

....

d   Describe what is meant by a dictionary data structure.

....

8 A vector, v, is described as [2,6]. Another vector, u, is described as [7,3]. Figure 35 shows a two-dimensional grid with an arrow representing vector v. The result of applying vector v from the origin is marked as position A. (AO1, AO2)   **20 marks**

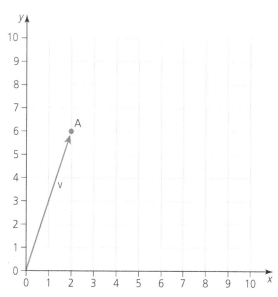

Figure 35

a   On the grid above, draw an arrow extending from position A using the vector u. Mark the ending position as B.

b   Calculate the result of u + v, showing your working.

....

....

c   On the grid above, draw an arrow from the origin to position B. Label the arrow as vector w. State the value of vector w.

....

d   State the name for the effect of adding two vectors.

....

e   State the name for the effect of multiplying a vector by a scalar.

....

f   Calculate the following scalar–vector products:

   i   3 * u

....

   ii   −2 * v

....

iii 0.5 * u

....

g Calculate the dot product of u and v.

....

....

h Describe a typical use for calculating a dot product of two vectors.

....

i Position Q is reached by moving according to vector q from the origin. Position R is reached by moving according to vector r from the origin. Position P lies directly on the line marking the shortest path between positions Q and R. It is estimated that the vector p can be described as 0.2q + 0.9r. Explain why this estimate must be incorrect.

....

....

....

Exam-style questions

(35)

9 A card game features a standard deck of 52 playing cards placed in a pile on top of each other. During the game players take cards from the top of the deck. When a player gives up a card it must also be placed at the top of the deck. (AO1, AO2, AO3)   **8 marks**

a Explain why a stack is a suitable data structure to represent the deck of cards in the game.

....

....

b State the name of the most suitable data structure to represent the deck of cards if returned cards were to be placed at the bottom of the deck.

....

The stack representing the deck has been implemented as a static array. Figure 36 shows part of the contents of the DeckStack array, The variable StackSize indicates how many cards are currently represented in the stack.

Index	[1]	[2]	[3]	...	[50]	[51]	[52]
Data	7 of diamonds	Queen of hearts	Six of clubs		9 of spades	Jack of diamonds	None

StackSize	51

Figure 36

c State the card that will be drawn next from the top of the stack.

....

d Write a pseudo-code algorithm to deal a card from the deck.

Your algorithm should output the value of the card that is to be dealt and make any required modifications to the contents of both the StackSize and DeckStack variables.

It should behave accordingly with any situation that might arise in the DeckStack array during the playing of the game.

....

....

....

....

....

....

....

....

....

....

10 Kieran is designing a rugby simulation. A rugby player, A, is located at position [3,2]. A second player, B, is located at [5,8] and is carrying the ball. B is running with a vector, v of [2,–5]. Figure 37 represents the current situation. (AO1, AO2)   **13 marks**

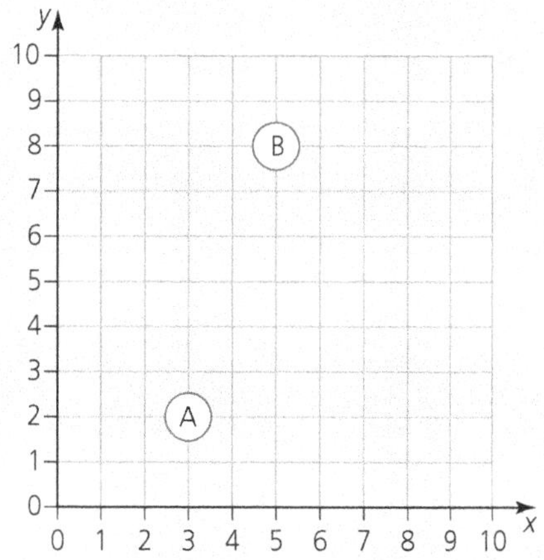

Figure 37

a Calculate B + v.

....

....

b On the grid above, plot the new position of B + v, label it C.

c Calculate the value of B + 4[v].

....

....

d The vector, u, describes the movement vector from C to A. State the value of u.

In the game of rugby the player with the ball can only throw it behind them. Kieran has identified the following algorithm to identify whether the player can legitimately throw the ball.

```
Value <- DotProduct(u,v)

IF Value = 0 THEN

    Allowed <- "Borderline"

ELSE IF Value < 0 THEN

    Allowed <- "Yes"

ELSE Allowed <- "No"

ENDIF
```

e Calculate the dot product of u and v and state the value of Allowed.

Kieran has developed the following algorithm:

```
Alpha <- 0.7

Beta <- 0.3

D <- [Alpha * B + Beta * C]
```

f Calculate the value of vector D.

g For each of the following values for Alpha and Beta, state whether the outcome of the algorithm above would be plotted somewhere on the line between B and C.

Alpha	Beta	Plotted on line BC?
0	1	
3	1	
0.1	0.9	
0.6	−0.4	
0.55	0.45	
−0.2	1.2	

Fundamentals of algorithms

Graph traversal, tree traversal and Reverse Polish Notation

Graph and tree traversal algorithms are typically recursive. You are not expected to memorise the algorithms themselves, but are expected to be able to trace programs to traverse graphs and trees, and to be familiar with the purposes of different algorithms.

Breadth-first search

Breadth-first algorithms involve moving to each node with a depth of 1 from the starting point, then moving to each node with a depth of 2, etc. This is best modelled using a queue data structure.

Depth-first search

Depth-first algorithms involve moving as far through the graph as possible and then taking one step back and repeating. This is best modelled using a stack data structure as you step back through the previously recorded nodes to find new paths.

Tree traversal

Tree-traversal algorithms can be pre-order, in-order or post-order. The key thing is to see whether the node is interrogated before, during or after the recursive calls.

---

1 **The graph in Figure 38 shows a series of cities and the flight paths between them. A programmer is writing a piece of software to find the route with the fewest steps between any two cities. They have decided to use a breadth-first search algorithm. (AO1, AO2)** `4 marks`

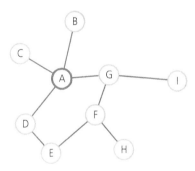

Figure 38

a **Describe a use for a depth-first search.**

....

b **State whether the programmer should use a stack data structure or a queue data structure when creating their breadth-first search algorithm, and why.**

....

....

c **Suggest one feature of a graph that may make it more complex to traverse than a tree.**

....

2 A competition is being held that tests both sailing skills and navigation. The challenge is to reach the finishing position via a series of waypoints. Waypoints can be visited in several possible orders and a graph to show the possible routes between waypoints is shown in Figure 39, with each ship starting at waypoint 0 and aiming to reach waypoint 7. It is not necessary to visit every waypoint. An algorithm for finding a route to waypoint 7 is shown in Figure 40. (AO1, AO2)

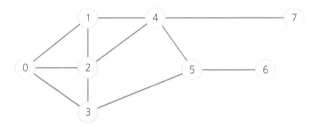

Figure 39

```
PROCEDURE DFS(V,T)

   Stack S ← [Empty]

   Discovered[0...7] ← [False]

   S.Push(V)

   WHILE S is Not Empty And V Not Equal T DO

        V ← S.Pop()

        IF Not Discovered[V] DO

           Discovered[V] ← True

           FOR each vertex U which is adjacent to V DO

              IF Not Discovered(U) AND Not S.Contains(U) DO

                 S.Push(U)

              ENDIF

           ENDFOR

        ENDIF

   ENDWHILE

ENDPROCEDURE
```

Figure 40

When looking for each vertex U which is adjacent to V, each vertex is scanned in ascending numerical order.

The algorithm in Figure 40 makes use of two subroutines, Pop() and Push(node).

Pop is a subroutine that deletes the top-most item from a stack and returns it to the main program.

Push is a subroutine that adds an item to the top of the stack.

Complete the trace table on page 36 to show how variables S, V, T and Discovered are updated when the algorithm is called using DFS(0,7).

The starting values for each variable have been entered for you.

The letter F has been used as an abbreviation for False. You should use the letter T as an abbreviation for True.

You may not need to complete every row.

V	T	Discovered								Stack	
		0	1	2	3	4	5	6	7	Bottom	Top
0	7	F	F	F	F	F	F	F	F	0	
0											

3 **a** Convert the following Reverse Polish Notation expressions into their equivalent infix expressions. (AO1, AO2)   **8 marks**

  i   19 3 +

....

  ii   12 4 * 6 +

....

b State whether a stack or a queue is most appropriate for evaluating an equation expressed in Reverse Polish Notation.

....

Figure 41 shows a tree that represents a mathematical equation.

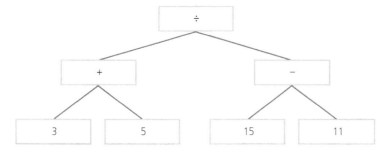

Figure 41

c State the output of traversing the tree in Figure 41 using in-order tree traversal.

....

d   State the output of traversing the tree in Figure 41 using post-order tree traversal.

....

e   State the name for the notation used in your response to part (c).

....

f   State the name for the notation used in your response to part (d).

....

g   Describe one problem with the equation as described in your response to part (c).

....

....

Exam-style questions

(30)

④ A model of an airline's flight routes is shown in Figure 42 using a graph data structure. An algorithm has been designed to find the shortest path between two vertices and this is shown in Figure 43. (AO1, AO2)

8 marks

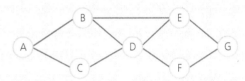

Figure 42

```
PROCEDURE ShortestPath(S, T)

   PutVertexIntoQueue(S)

   Discovered[S] ← True

   Found ← False

   WHILE Queue Not Empty AND Found = False DO

      V = GetVertexFromQueue()

      FOR each vertex U adjacent to V DO

         IF Discovered[U] = False AND Found = False AND S.Contains(U) = False THEN

            PutVertexIntoQueue(U)

            Discovered[U] ← True

            Parent[U] ← V

            IF U = T THEN

               Found ← True

            ENDIF

         ENDIF

      ENDFOR

   ENDWHILE

ENDPROCEDURE
```

Figure 43

The algorithm shown in Figure 43 contains two subroutines, GetVertexFromQueue() and PutVertexIntoQueue(vertex).

The subroutine GetVertexFromQueue will delete the item at the front of the queue and return it.

The subroutine PutVertexIntoQueue will insert the passed item at the back of the queue.

a   Complete the trace table below to show what happens to each variable when the algorithm in Figure 43 is run using the graph in Figure 42.

The starting values for each variables have been entered for you.

The letter F has been used as an abbreviation for False. You should use the letter T as an abbreviation for True.

You may not need to complete every row.

S	T	F	Queue		Discovered							Parent							
			Front	Back	A	B	C	D	E	F	G	A	B	C	D	E	F	G	V
A	G	F	A		T	F	F	F	F	F	F								
			Empty																A

b   State the type of graph traversal used in the algorithm shown in Figure 43.

....

c   Name and describe a typical use for another type of graph traversal algorithm.

....

....

5   The diagram in Figure 44 shows the names of various operating systems. (AO1, AO2)   4 marks

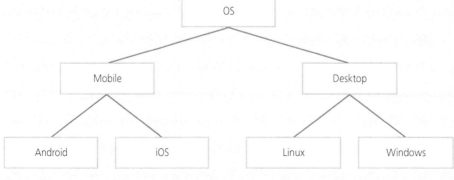

Figure 44

a State the most appropriate name for the data structure shown in Figure 44.

....

An algorithm has been designed that will output the items in the data structure shown in Figure 44 in the following order: OS, Mobile, Android, iOS, Desktop, Linux, Windows.

b State the name of the standard traversal algorithm that will produce this output.

....

c Some graphs require a more complex traversal algorithm. Suggest one reason why this might be the case.

....

6 Figure 45 shows a data structure containing a mathematical equation. An algorithm for traversing the data structure is shown in Figure 46. (AO1, AO2)

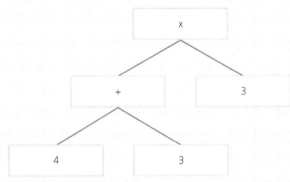

Figure 45

```
PROCEDURE Traverse(node)
    IF Exists(node.left) THEN
       Traverse(node.left)
    ENDIF
    OUTPUT node
    IF Exists(node.right) THEN
       Traverse(node.right)
    ENDIF
ENDPROCEDURE
```

Figure 46

a State the values which are output when running the algorithm in Figure 46 with the data structure shown in Figure 45.

....

....

b Explain why a computer may struggle to process the equation in the format shown in your response to part (a).

....

....

c State the name of the traversal algorithm shown in Figure 46.

....

d State the name of the traversal algorithm that will output the equation in Reverse Polish Notation.

....

e Convert the following Reverse Polish Notation equations to infix notation.

i 2 6*

....

ii 4 3 + 2 −

....

iii 7 4 * 2 + 10 ÷

....

Searching, sorting and optimisation algorithms

You are not expected to memorise the algorithms in this topic completely, though you are expected to be able to trace the algorithms, identify them and analyse their time complexity (see regular and context-free languages).

Searching

Searching algorithms are generally relatively simple to follow and relatively quick to execute. The most basic example is the linear search, in which the items are placed in a line and the program will simply check each value one at a time until it finds the one it needs. This is very simple, but inefficient. A binary search checks the item in the middle of a list, meaning half of the values can be discarded each time. A binary search is only possible if the list is already sorted.

Sorting

Sorting algorithms are more complex than searching algorithms, but make searching easier as a binary search can be implemented. A bubble sort is relatively simple to follow but very inefficient, with multiple passes required to ensure that the list is fully in order. A merge sort is much more efficient, but also much more complicated.

Optimisation algorithms

Dijkstra's algorithm is the only optimisation algorithm you are specifically expected to have met. You are not expected to recall the exact steps or analyse the time complexity, but you should be able to trace the algorithm and understand that it is used to find the shortest route between two nodes in a weighted graph.

1 The table below shows the contents of an array, Fruits, which has been sorted in alphabetical order. (AO1, AO2)

Figure 47 shows an algorithm that has been designed to allow a user to find an item in the list.

[0]	[1]	[2]	[3]	[4]	[5]	[6]	[7]
Apple	Banana	Lime	Kiwi	Mango	Orange	Pear	Tomato

```
PROCEDURE BinarySearch(Array, SearchWord)

    First ← 0

    Last ← Array.Length - 1

    Found ← False

    Failed ← False

    WHILE Not Failed AND Not Found DO

        Midpoint ← (First + Last) DIV 2

        IF List[Midpoint] = SearchWord THEN

            Found ← True

        ELSE

            IF First >= Last THEN

                Failed ← True

            ELSE

                IF List[Midpoint] > SearchWord THEN

                    Last ← Midpoint - 1

                ELSE

                    First ← Midpoint + 1

                ENDIF

            ENDIF

        ENDIF

    ENDWHILE

    IF Found = True THEN

        OUTPUT "Found it!"

    ELSE

        OUTPUT "Sorry, not found"

    ENDIF

ENDPROCEDURE
```

Figure 47

a   Complete the trace table below to show the result of running BinarySearch(Fruit, "Orange").
(Note: It may not be necessary to complete all the rows.)

First	Last	Found	Failed	Midpoint	Output
0	7	False	False	3	

b   Complete the trace table below to show the result of running BinarySearch(Fruit, "Grape").

First	Last	Found	Failed	Midpoint	Output
0	7	False	False	3	

c   State one advantage of use a binary search instead of a linear search.

....

d   State one disadvantage of using a binary search instead of a linear search.

....

2   The table below shows the contents of an array of integers, Scores. (AO1, AO2, AO3)   8 marks

[0]	[1]	[2]	[3]	[4]	[5]
3	12	5	9	4	1

Figure 48 shows an incomplete attempt at creating a bubble sort algorithm.

```
FUNCTION Bubble (Scores)

    Length ← Scores.LENGTH()

    ..................................................................

        FOR j ← 1 TO LENGTH − 1 DO

            IF Scores[j] > Scores [j+1] THEN

                ..................................................................

                ..................................................................

                ..................................................................

            ENDIF

        ENDFOR

    ..................................................................

    RETURN Scores

ENDFUNCTION
```

Figure 48

a   Complete the pseudocode algorithm in Figure 48.

b   Complete the table below to show the values in the array after one pass of the completed bubble sort from Figure 48.

[0]	[1]	[2]	[3]	[4]	[5]

c   The bubble sort algorithm is considered to be inefficient.

   i   Describe one example of inefficiency in the bubble sort algorithm.

....

ii Explain how the algorithm could be changed to reduce the number of comparisons required.

....

....

....

3 Figure 49 shows a weighted graph representing the time it takes to travel between junctions in a road system. Figure 50 shows a pseudo-code algorithm that can be used on this graph. (AO1, AO2)   **8 marks**

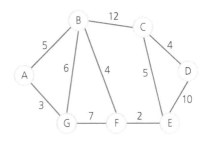

Figure 49

```
FUNCTION Dijkstra(Graph, Start)
    Q ← the set of all nodes in Graph
    FOR each Vertex V in Graph
        Dist[V] ← Infinity
        Previous[V] ← Empty
    ENDFOR
    Dist[Start] ← 0
    WHILE Q Not Empty
        U ← node in Q with smallest Dist[ ]
        Remove U from Q
        FOR each neighbour V of U in Q
            Alt ← Dist[U] + DistBetween(U, V)
            IF Alt < Dist[V]
                Dist[V] ← Alt
                Previous[V] ← U
            ENDIF
        ENDFOR
    ENDWHILE
    Return Previous
ENDFUNCTION
```

Figure 50

a   Complete the trace table below to show the result of running the algorithm in Figure 50 on the graph in Figure 49, starting from Node A.

U	V	Alt	Dist							Previous							Q
			A	B	C	D	E	F	G	A	B	C	D	E	F	G	ABCDEFG
			∞	∞	∞	∞	∞	∞	∞								
			0														
A																	BCDEFG

b   Complete the table below to show the values of the array that is returned.

A	B	C	D	E	F	G

c   State the shortest path from node A to node D.

....

Exam-style questions

4   Anika is writing a piece of code to search for an item in the array, Subjects, shown below. (AO1, AO2)   **7 marks**

Subjects					
[0]	[1]	[2]	[3]	[4]	[5]
German	Computer Science	Maths	Media Studies	English	Art

a   State the reason why a binary search would not be suitable for searching this array.

....

b   State the name of the searching algorithm that could be used to search this array.

....

c State the values that would be compared if Anika was to search the array for 'Media Studies' using this algorithm.

....

Figure 51 shows an incomplete binary tree that is designed to hold the same values as the array, Subjects.

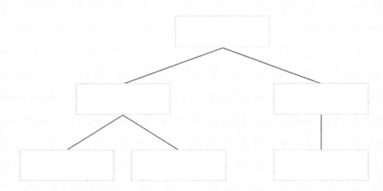

Figure 51

d Using Subjects[0] as the root, insert the items from the array Subjects into the binary tree above.

e State the maximum number of comparisons needed to find an item in the binary tree using a binary tree search.

....

f Describe the sequence of comparisons if Anika performed a binary tree search for 'History' using the completed binary tree from Figure 51.

....

....

5 Figure 52 shows a pseudo-code algorithm. (AO1, AO2)　　　　　　　　　　　　　　　6 marks

```
FUNCTION Hodder(Array)

    Length ← Array.LENGTH()

    FOR i ← 1 TO LENGTH - 1 DO

        FOR j ← 1 TO LENGTH - 1 DO

            IF Array[j] > Array [j+1] THEN

                Temp ← Array[j]

                Array[j] ← Array[j+1]

                Array [j+1] ← Temp

            ENDIF

        ENDFOR

    ENDFOR

    RETURN Array

ENDFUNCTION
```

Figure 52

a Complete the trace table below to show the result of running one pass through the outer loop of the algorithm shown in Figure 52. The initial array values have already been completed for you.

i	j	Temp	Array					
			[1]	[2]	[3]	[4]	[5]	[6]
			216	124	365	211	53	129

b State the name of the algorithm shown in Figure 52.

....

c Complete the table below to show the values of the array after a total of three passes through the algorithm shown in Figure 52.

....

....

....

[1]	[2]	[3]	[4]	[5]	[6]

Theory of computation

Abstraction and automation

Syllogisms

Logical syllogisms often form part of Paper 1 and it can sometimes help to draw a Venn diagram to visualise the possibilities. The wording is often odd-looking in order to help you focus on the abstract logic and not your existing knowledge of that topic.

Abstraction

Abstraction essentially focuses on reducing complexity. This could be through ignoring unimportant details or through the hiding of complexity. Functions and procedures can be used to hide the finer details of an algorithm and data structures can also be used to hide complexity.

1. **For each of the following questions, highlight or circle ONE correct response in each case. (AO2)**

   a **Statements**

      No swimming teachers can play. Some swimming teachers are athletes.

      Conclusions

      I **Gymnastics athletes can play.**
      II **Some athletes can play.**

         i **Only conclusion I follows**

         ii **Only conclusion II follows**

         iii **Either I or II follows**

         iv **Neither I nor II follows**

         v **Both I and II follow**

   b **Statements**

      All hats are cakes. All coats are cakes.

      Conclusions

      I **Some coats are hats.**
      II **No coat is a hat.**

         i **Only conclusion I follows**

         ii **Only conclusion II follows**

         iii **Either I or II follows**

         iv **Neither I nor II follows**

         v **Both I and II follow**

Regular and context-free languages

Regular languages

English and other human languages are imprecise and full of ambiguities. Finite state machines, regular expressions and Backus Naur Form are all methods of representing languages that are more easily and reliably processed by computers. Being able to read and apply your knowledge of these languages is essential.

1 The finite state machine (FSM) represented as a state transition diagram in Figure 53 recognises a language with an alphabet of a, b and c.

8 marks

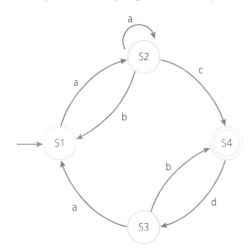

Figure 53

a In the table below, indicate whether each input string is accepted or not accepted by the FSM in Figure 53.

 ▪ If an input string is accepted write YES.
 ▪ If an input string is not accepted write NO.

Input string	Accepted by FSM?
aaaac	
abc	
abaacdb	

b In words, describe the language (set of strings) that is accepted by the FSM in Figure 53.

....

....

....

2 Figure 54 show a list of different sets. (AO1, AO2)

10 marks

```
S1  = {m,n,p}
S2  = {m,n}
S3  = {1,2,3,4}
S4  = {0,1,2}
S5  = {m,mn,mp}
S6  = {r}
S7  = {m,n,p}
S8  = {∅}
S9  = {p}
S10 = ℕ
```

Figure 54

a State the name of a set listed in Figure 54 that is empty.

....

b   Describe, using example sets listed in Figure 54, the difference between a subset of S1 and a proper subset of S1.

....

....

....

....

c   State the name of a set listed in Figure 54 that has the same cardinality as S1 but is not a subset of S1.

....

d   State the name of a set listed in Figure 54 that is infinite.

....

e   Describe how the set S2 could be created using a difference operation together with two other sets listed in Figure 54.

....

f   What is the cartesian product of sets S6 and S2?

....

3 An example of a regular expression is (a|b)c*d+. (AO1, AO2)   **9 marks**

a   What is the meaning of the symbol '*'?

....

b   What is the meaning of the symbol '+'?

....

c   For each of the following inputs, state whether it would or would not be accepted.

Input	Accepted
acd	
bcccccd	
acccc	
abccd	

d   The regular expression has been rewritten as (a|b)?c*d+. What is the meaning of the symbol '?'?

....

e   Complete the table below.

Input	Regular expression	Accepted
color	colou*r	
colour	colou*r	
colouur	colou*r	
color	colou?r	
colour	colou?r	
colouur	colou?r	

4 Dogs registered to the Kennel Club are to be stored in a file according to a set of rules which are described in Backus Naur Form in Figure 55. (AO1, AO2)   **5 marks**

An underscore (_) represents a space.

The rules are not case sensitive.

```
1  <KCrecord>  ::=  <KCname> _ - _ <breed>

2  <KCname>  ::=  <kennelname> _ <dogname>

3  <breed>  ::=  <name>  |  <breed> _ <name>

4  <dogname>  ::=  <name>  |  <dogname> _ <name>

5  <kennelname>  ::=  <name>  |  <kennelname> _ <name>

6  <name>  ::=  <letter>  |  <name> _ <letter>

7  <letter>  ::=  a|b|c|d|e|f|g|h|i|j|k|l|m|n|o|p|
                  q|r|s|t|u|v|w|x|y|z
```

Figure 55

a State the base case for rule 4.

....

b Explain what is meant by a base case in a BNF rule.

....

....

c For each of the rules in Figure 55, state whether it could be defined using a regular expression.

Rule number	Could be defined using a regular expression
1	
2	
3	
4	
5	
6	
7	

d State a simple rule for recognising which of the BNF rules in Figure 55 could or could not be defined using a regular expression.

....

e State which of the following KCrecords would be accepted using the BNF rules shown in Figure 55.

KCname	Accepted
Ypres Prince Hektor - Miniature Schnauzer	
Spike - Daschund	
Shergar Bernhard T Doberman	
Kilcullen Rocky - Spaniel	

Exam-style questions

5 The finite state machine in Figure 56 represents the possible states of a hairdryer with two switches – one for hot or cold, and one for off or on. (AO1, AO2)

 4 marks

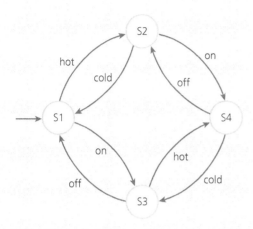

Figure 56

a If the FSM in Figure 56 reaches S4, what does it mean?

....

b An FSM can be represented as a state transition diagram or as a state transition table. Complete the table below.

Original state	Input	New state
S2		
S2		
S4		
S4		

c If the hairdryer was upgraded to one with three power settings – off, slow and fast – what is the minimum number of states that would need to be added to the FSM in Figure 56?

....

6 The finite state machine represented in Figure 57 recognises an input of a binary number.

 4 marks

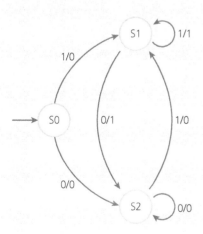

Figure 57

a State the corresponding outputs for each of the following inputs. All inputs start with the most significant bit (MSB).

Input	Output
0110	
1001 1010	
0101 0101	

b State the purpose of this finite state machine.

....

7 UK number plates issued between 1983 and 2001 follow the following format.
(AO1, AO2)

6 marks

- One letter
- Between 1 and 3 digits
- Three letters

For example, R615 XWG

- The first letter denotes the year the vehicle was first registered.
- Kit cars and rebuilt cars are assigned a Q as their first letter.
- No cars are assigned a first letter of I, O, U or Z.

Number plates are often written with added spacing to improve readability, but they are stored without any white space.

A state transition diagram for a finite state machine to represent the rules above is shown in Figure 58.

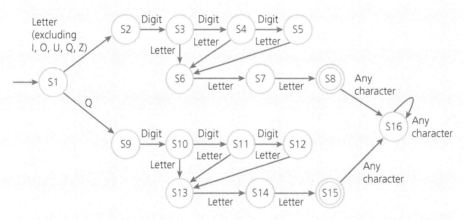

Figure 58

a For each of the following registration plates, state whether it would be accepted by writing YES or NO in the table below.

Registration plate	Accepted
RRM55G	
G55RRM	
F612RT	
L731PRMF	
T527QHI	

b If the FSM shown in Figure 58 ends at S15, what does this mean?

....

....

A different method of checking the validity of a registration plate checks that it fits one of the following rules:

Rule 1

- One letter
- Followed by 1–3 digits
- Followed by 3 letters

Rule 2

- Two letters
- Followed by 2 digits
- Followed by 3 letters

c Using the character \d to represent any digit and \l to represent any letter, write a regular expression that represents a valid number plate using the rules above.

....

....

....

Classification of algorithms

Time complexity

Commonly referred to as Big-O (or Big-Oh), time complexity is crucial in measuring the efficiency of algorithms in terms of the time it takes to work with large data sets. It is important to be aware that Big-O refers to time complexity, but space complexity (the amount of memory needed) is also a significant factor.

1 Below is an unordered list of possible time complexities. (AO1, AO2) **7 marks**

- $O(n)$
- $O(n^2)$
- $O(\log n)$
- $O(n \log n)$
- $O(n!)$
- $O(k^n)$
- $O(1)$

State the time complexity for each of the following standard algorithms.

a Linear search

....

b Bubble sort

....

c Binary search

....

d Merge sort

....

e Finding the first item in a list

....

f Creating every possible order of letters from a word

....

g Binary tree search

....

2 Time complexity is a measure of the efficiency of an algorithm. (AO1, AO2) **14 marks**

a Other than time complexity, state one other measure of the efficiency of an algorithm.

....

b State the time complexity of an algorithm that is tractable.

....

c State the time complexity of an algorithm that is intractable.

....

d Explain what is meant by an intractable problem.

....

....

e Suggest two possible strategies for a programmer tasked with solving an intractable problem.

....

....

f For each of the following algorithms, state whether it is tractable, intractable or unsolvable.

Problem	Classification
Finding an item in an ordered list	
The travelling salesman problem	
The Halting problem	
Sorting a list into order	
Checking every combination of letters in an anagram solver	

g Explain what is meant by the Halting problem.

....

....

A model of computation

Turing machines

A Turing machine can be thought of as a simple computer with one, fixed program and an infinitely long tape on which values can be stored. A Turing machine provides a formal model of computation and a programming language is considered to be Turing-complete if it can be used to simulate any Turing machine.

1 A Turing machine is a device capable of carrying out an algorithm. (AO1, AO2)   **5 marks**

a Complete the table below by writing True or False next to each statement.

Statement	True or False
If an algorithm exists to solve a computational problem then a Turing machine can be designed to solve the problem	
If no algorithm exists to solve a computational problem then no Turing machine can be designed to solve it	
Some computational problems which can be solved by an algorithm cannot be solved by a Turing machine	

A Turing machine has states S_0, S_1 and S_F. S_0 is the start state and S_F is the stop state.

The machine stores data on a single tape which is infinitely long in one direction. The machine's alphabet is 0, 1 and □, where □ represents a blank cell.

The Turing machine's transition function is defined by:

$$\delta(S_0,0) = (S_0,0,\rightarrow) \qquad \delta(S_1,0) = (S_0,1,\rightarrow)$$
$$\delta(S_0,1) = (S_1,0,\rightarrow) \qquad \delta(S_1,1) = (S_1,1,\rightarrow)$$
$$\delta(S_0,\square) = (S_F,\square,\leftarrow) \qquad \delta(S_1,\square) = (S_F,\square,\leftarrow)$$

b Trace the computation of the Turing machine, using the transition function δ.

Show the contents of the tape, the current position of the read/write head and the current state as the input symbols are processed.

The initial configuration of the machine has been completed for you in step 1.

Figure 59

c State the purpose of the Turing machine described in part (b).

....

Exam-style questions   ⏱ 10

2 A Turing machine has been designed to have states S_0, S_1 and S_F. S_0 is the start state and S_F is the stop state. (AO1, AO2)   **7 marks**

The machine stores data on a single tape which is infinitely long in one direction. The machine's alphabet is 0, 1 and □, where □ represents a blank cell.

The Turing machine's transition function is defined by:

$$\delta(S_0,0) = (S_0,0,\leftarrow) \qquad \delta(S_1,0) = (S_1,1,\leftarrow)$$
$$\delta(S_0,1) = (S_1,1,\leftarrow) \qquad \delta(S_1,1) = (S_1,0,\leftarrow)$$
$$\delta(S_0,\square) = (S_F,\square,\rightarrow) \qquad \delta(S_1,\square) = (S_F,\square,\rightarrow)$$

a Trace the computation of the Turing machine, using the transition function δ.

Show the contents of the tape, the current position of the read/write head and the current state as the input symbols are processed.

The initial configuration of the machine has been completed for you in step 1.

1. ··· | 0 | 0 | 1 | 1 | 0 | 1 | 0 | 0 | | S_0
State
↑

2. ··· | | | | | | | | | | State

3. ··· | | | | | | | | | | State

4. ··· | | | | | | | | | | State

5. ··· | | | | | | | | | | State

6. ··· | | | | | | | | | | State

7. ··· | | | | | | | | | | State

8. ··· | | | | | | | | | | State

9. ··· | | | | | | | | | | State

10. ··· | | | | | | | | | | State

Figure 60

b State the purpose of the Turing machine described in this question.

....

Cover photo: jim/Adobe Stock

Hachette UK's policy is to use papers that are natural, renewable and recyclable products and made from wood grown in well-managed forests and other controlled sources. The logging and manufacturing processes are expected to conform to the environmental regulations of the country of origin.

Orders: please contact Hachette UK Distribution, Hely Hutchinson Centre, Milton Road, Didcot, Oxfordshire, OX11 7HH

Telephone: 01235 827827

Email: education@hachette.co.uk

Lines are open from 9 a.m. to 5 p.m., Monday to Friday. You can also order through our website: www.hoddereducation.co.uk

ISBN: 978-1-5104-3701-2

© Mark Clarkson 2019

First published in 2019 by
Hodder Education,
An Hachette UK company
Carmelite House
50 Victoria Embankment
London EC4Y 0DZ

www.hoddereducation.co.uk

Impression number   10 9 8 7 6 5

Year   2023

All rights reserved. Apart from any use permitted under UK copyright law, no part of this publication may be reproduced or transmitted in any form or by any means, electronic or mechanical, including photocopying and recording, or held within any information storage and retrieval system, without permission in writing from the publisher or under licence from the Copyright Licensing Agency Limited. Further details of such licences (for reprographic reproduction) may be obtained from the Copyright Licensing Agency Limited, www.cla.co.uk

Typeset by Aptara, India

Printed by Ashford Colour Press Ltd.

A catalogue record for this title is available from the British Library.

HODDER EDUCATION

t: 01235 827827
e: education@hachette.co.uk
w: hoddereducation.co.uk

ISBN 978-1-5104-3701-2

MIX
Paper | Supporting responsible forestry
FSC™ C104740